Inse

by Maryellen Gregoire

Consultant:
Adria F. Klein, Ph.D.
California State University, San Bernardino

capstone
classroom
Heinemann Raintree • Red Brick Learning
division of Capstone

A ladybug is an insect.

A grasshopper is an insect.

An ant is an insect.

A bee is an insect.

A butterfly is an insect.

A dragonfly is an insect.

A cockroach is an insect.

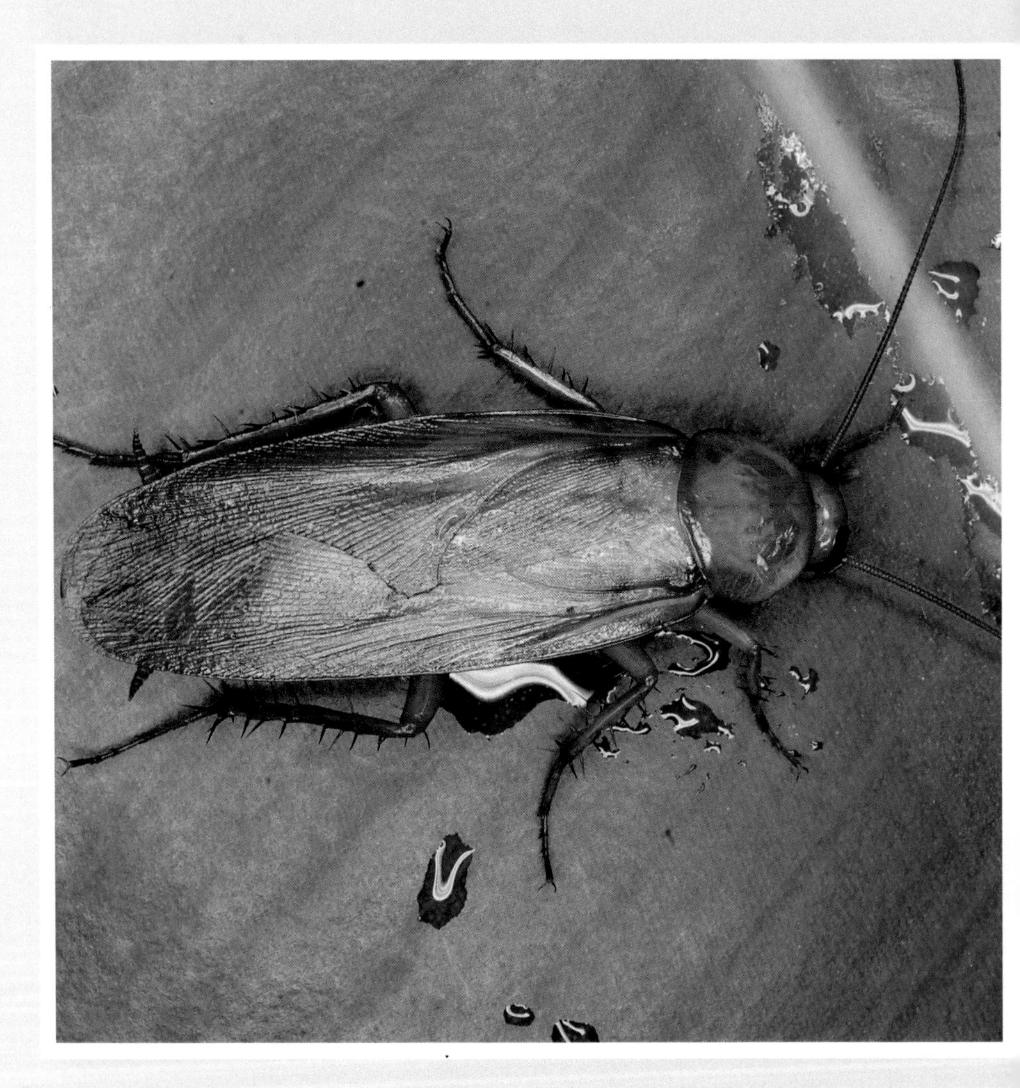